A Science Comic of Urban Metro Structure

Hehua Zhu

A Science Comic of Urban Metro Structure

Performance Evolution and Sensing Control

Hehua Zhu
Department of Geotechnical Engineering,
    College of Civil Engineering
Tongji University
Shanghai, China

ISBN 978-981-13-4456-5   ISBN 978-981-13-0580-1   (eBook)
https://doi.org/10.1007/978-981-13-0580-1

Jointly published with Tongji University Press, Shanghai, China

The print edition is not for sale in China Mainland. Customers from China Mainland please order the print book from: Tongji University Press.

This Springer imprint is published by the registered company Springer Nature Singapore Pte Ltd.
The registered company address is: 152 Beach Road, #21-01/04 Gateway East, Singapore 189721, Singapore

Preface

Due to the characteristics of high speed, great transportation capacity, safety, comfort, convenience, etc., urban rail transit has become one of the core members of urban public transportation network. Among the different parts of urban rail transit, the underground part is called underground rail, or metro for short. So far, the daily passenger volume is more than 10,000,000 on weekdays in Shanghai, which is 49% of the total amount of public transportation in the whole city according to the statistical data. However, when you are travelling through the metro that is tens of meters under the ground, have you ever thought about the "safety"? How much do you know about this common but significantly important urban transportation? What is the tunnel that bears the operation of metro? Does the tunnel deteriorate like the human being? How can we know its condition? How to deal with the distress that occurs in the tunnel? Therefore, it is very important to answer those questions scientifically for every urban resident.

In order to solve those questions above to some extent, the National Basic Research and Development Program (973 program) "Fundamental Theory for The Performance Evolution and Sensing-Control of Urban Metro Structures" was conducted by the national research groups from Tongji University, Huazhong University of Science and Technology, Central South University, South China University of Technology, Nanjing University of Technology, and Shanghai Shen Tong Metro Group Co., Ltd. from November 2011 to August 2016. The project includes the research in the following six areas: (1) life-cycle performance evolution mechanism of underground structural material under dynamic service environment, (2) coupled mechanism of underground structure and environment, (3) smart sensing theory and method of super long and linear underground structure, (4) health diagnosis and service performance prediction of underground structure under the dynamic space-time environment, (5) structural self-healing and reinforcement theory under the groundwater environment, and (6) digital maintenance and control system for underground structure. After five years of hard work, the research group has successfully solved the difficult problems and challenges. Fruitful achievements have been made and the concluding report will answer the questions above. On the basis of the research achievements, we were thinking and planning to draw science comic so that the program achievements can be presented in an intuitive and vivid way, and the public can get to know the safety and health problems of metros in a simple way.

The science comic consist of the life-cycle period of a metro including the birth (overview), sickness (structure condition and distress), medical records (digital information archive), experts' diagnoses (structure sensing), consultation (structure health condition evaluation and prediction), and treatment (structure repair, reinforcement, and control). The content includes seven parts. The theme of the first part is "Better metro, better city", illustrating the importance of metro transportation on modern cities. The second and third parts describe "What does a metro tunnel look like" and "Metro tunnel can get sick", which is the research of Area 1 and 2. The fourth part introduces "Digital information archive of metro tunnel", which is the research of Area 6. The fifth, sixth, and seventh parts introduce the process of "Experts' diagnosis, consultation and treatment", which is the research of Areas 3, 4, and 5, respectively. Finally, the science comic will show us that the metro will serve and accompany us for one hundred years through the endless efforts and industrious guard.

The science comic is fully supported by the entire research team of the 973 program. We especially appreciate the work of Weiqing Liu, Limin Peng, Hongwei Huang, Hongping Zhu, Bo Wu, and Tinghui Bai who are in charge of the branch projects. Sincere thanks go to the key members and participants of each project for their efforts, including Xiaojun Li, Yongchang Cai, Shuguang Wang, Chenghua Shi, Hui Luo, Qing Chen, Yulong Zuo, Jiande Han, Shun Weng, Mingfeng Lei, Fei Wang, Yichao Ye, Yuexiang Lin, Xinyao Nie, Wuzhou Zhai, Shuo Zhang, Yuechun Luo, Jianbo Zang, Xueqin Chen, Xiaodong Lin, Nan Chen, Lianyang Zhang, Xiaoying Zhuang, et al. Many domestic and abroad technical reports and literature were referred during the process of completing the 973 program. The corresponding authors and research institutions are appreciated.

Ningxia Yang, Yi Hu, et al. from Tongji University Press are also appreciated for their support on the publication of the science comic.

This is our first attempt to present the research outcomes of the 973 program to the readers through the science comic. Since there is no precedent to follow and the time is limited, it is inevitable that there are parts which have not been well considered in the comic. We sincerely welcome the comments and corrections from the readers.

Shanghai, China   Hehua Zhu
June 2016   Professor in Underground Engineering, Principal
                        Scientist of the National 973 Program

Contents

1

Better Metro, Better City

The underground part of urban rail transit is called underground rail, or metro for short. As a common commuting transportation vehicle, metro runs tirelessly under the ground, conveying the urban passengers every day. Metro is easily accessible, fast, convenient, and with no congestion. It has changed our lifestyle without notice and become a habit for the urban resident to take the metro. However, how much do you know about the familiar metro? What does the tunnel structure look like? Is it safe? How would be its health condition? How to deal with the distress? Let's start learning about the metro system from a different perspective.

© Springer Nature Singapore Pte Ltd. and Tongji University Press 2019
H. Zhu, *A Science Comic of Urban Metro Structure*,
https://doi.org/10.1007/978-981-13-0580-1_1

As a common commuting transportation vehicle, metro links every corner in the city.

The metro system has greatly changed the pace and patterns of our lives. People can travel to their destinations free from the worries of congestions.

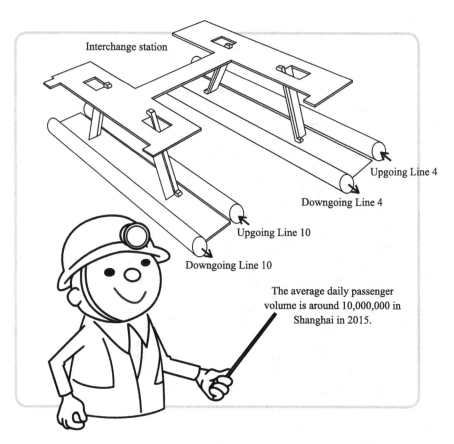

Interchange station

Upgoing Line 4

Downgoing Line 4

Upgoing Line 10

Downgoing Line 10

The average daily passenger volume is around 10,000,000 in Shanghai in 2015.

Metro trains are running tirelessly under the urban ground. For example, the average daily passenger volume is around 10,000,000 in Shanghai in 2015.

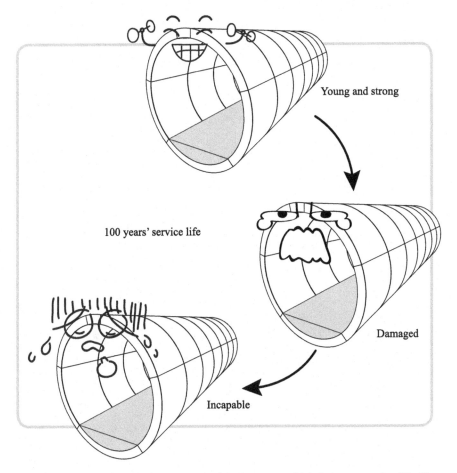

Young and strong

100 years' service life

Damaged

Incapable

The metros are supported by the tunnel structures, which have a service life like a human. Due to the heavy workloads and the complicated working environment, the tunnel structures could deteriorate from a young and healthy state to a damaged state and later enters its decaying stage, just as the aging process of a human.

2

What Does a Metro Tunnel Look like?

A metro tunnel is the section tunnel between two adjacent stations. It is a super long and linear structure connected by the lining ring extending in the longitudinal direction. The lining ring is assembled with the shield segments. The shield segment is a reinforced concrete structure, whose steel bar is like the bone of a human being and concrete is like the muscle on the bones. The steel bar and concrete bear the external loadings together. The shield segment is connected through the circumferential and longitudinal bolts in the circumferential and longitudinal directions, respectively. The bolt is like the joint of a human being. Therefore, the metro tunnel is composed of the shield segments, bolts, waterproof materials material, and some other components.

© Springer Nature Singapore Pte Ltd. and Tongji University Press 2019
H. Zhu, *A Science Comic of Urban Metro Structure*,
https://doi.org/10.1007/978-981-13-0580-1_2

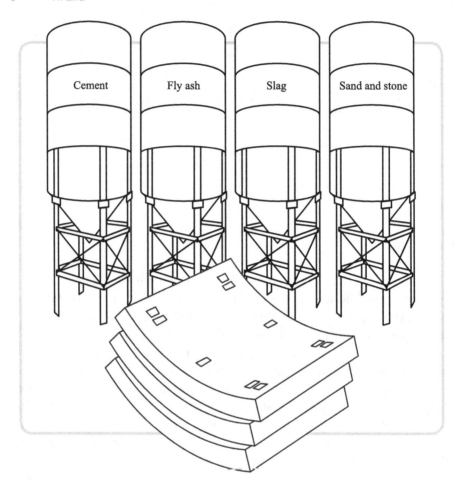

The raw materials, including the cement, fly ash, slag, sand, and stone, are utilized to produce the concrete of the shield segment.

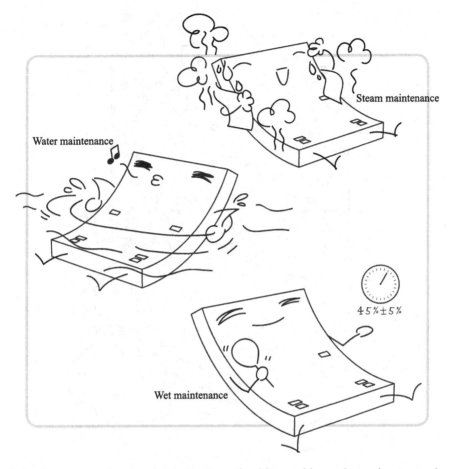

To let the concrete develop fast in its strength with a rapid speed, step-by-step maintenance processes are employed, including steam maintenance, water maintenance, and wet maintenance.

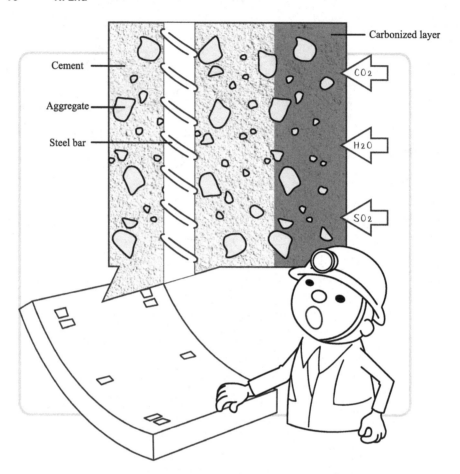

Carbonized layer

Cement

Aggregate

Steel bar

CO_2

H_2O

SO_2

Apart from the concrete, which may be deteriorated by the carbonization and sulfate attack, the steel bars are also needed to produce the reinforced concrete structure of shield segment.

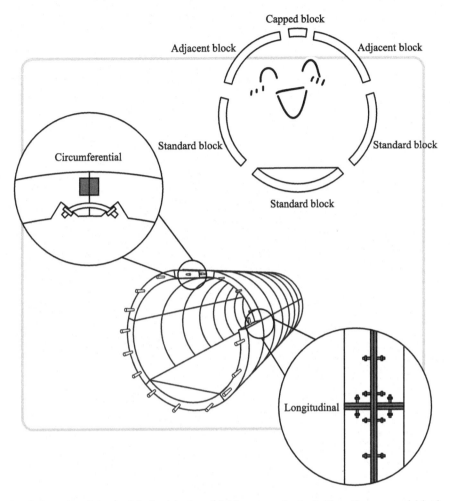

The lining ring is assembled with the shield segments, including the capped block, adjacent block, and standard block. The shield segment is connected through the circumferential and longitudinal bolts in the circumferential and longitudinal directions.

3

Metro Tunnel Can Get Sick

Metro tunnel serves in the complex and diverse underground environment, bearing the strata pressure, dynamic traffic loading, the disturbance caused by the projects in the neighborhood, etc. and interacting with the geotechnical media, erosive ions in the air, underground water, etc. With the fluctuation of air temperature and humidity, groundwater seepage and ion migration, the underground environment is changing all the time. The coupling effect between the metro tunnel and environment will lead to the performance change of the tunnel structure. Under such a dynamic spatial-temporal environment, it is inevitable that the distress will occur in the tunnel and the serviceability will be challenged.

© Springer Nature Singapore Pte Ltd. and Tongji University Press 2019
H. Zhu, *A Science Comic of Urban Metro Structure*,
https://doi.org/10.1007/978-981-13-0580-1_3

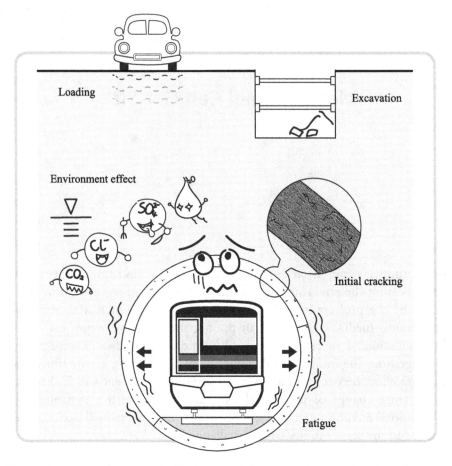

There are many environmental effects around the tunnel structures, such as the excavation, fatigue, and chloridion.

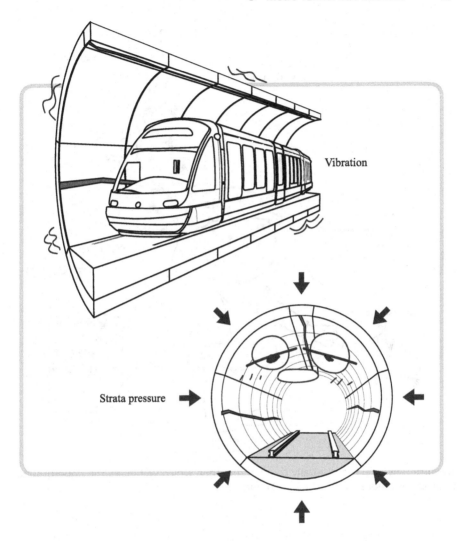

Vibration

Strata pressure

The tunnel structures must be strong enough to bear all kinds of effects. Or the metro tunnel will be in danger.

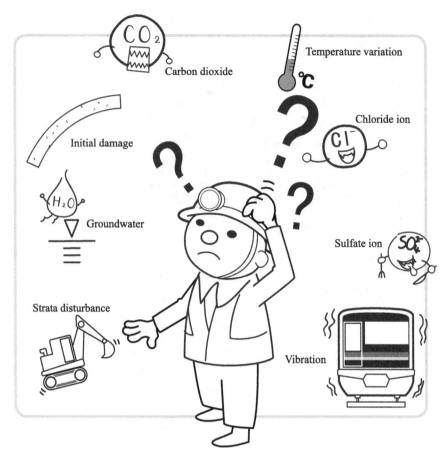

The different effects are usually coupled together, which will make the underground environment more complicated and dangerous.

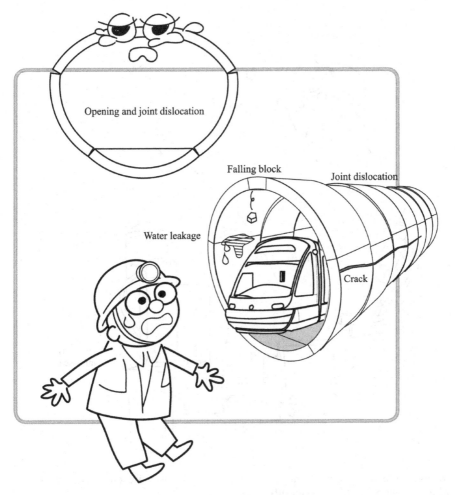

These environmental effects may lead to the opening and joint dislocation, falling block, water leakage, and crack, etc. To certain extend, the tunnel structures are ill when one of the mentioned phenomena happens.

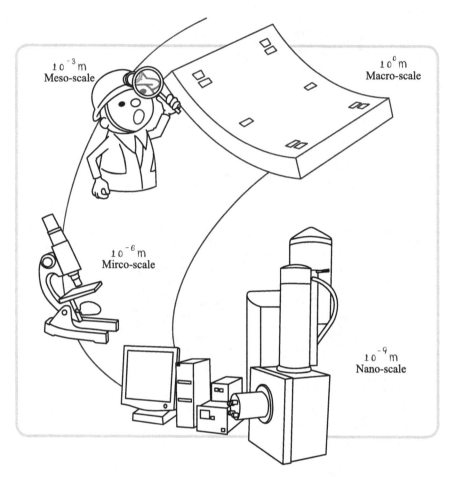

To know the deterioration mechanism, different devises are employed to observe what happened at different length scales.

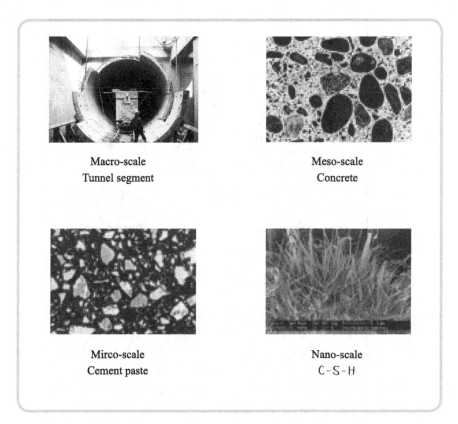

Macro-scale
Tunnel segment

Meso-scale
Concrete

Mirco-scale
Cement paste

Nano-scale
C-S-H

There are different length scale pore structures in the medium, which provide the intrusion path for the ions and lead to the illness of the segment.

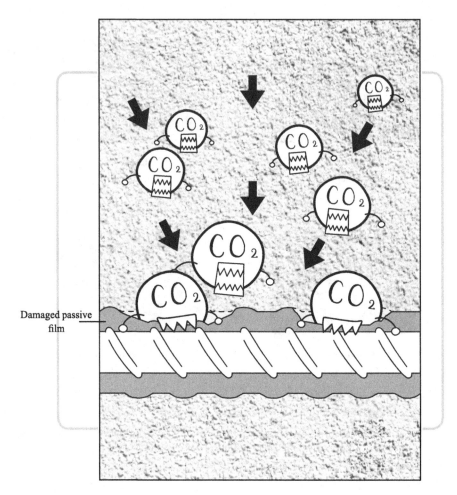

Damaged passive film

Through these paths, the CO_2 can lead to the carbonation of segment, which will damage the protective film on the steel bar.

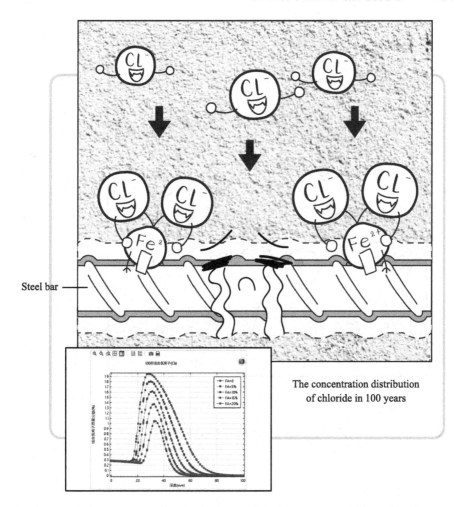

Steel bar

The concentration distribution of chloride in 100 years

Similarly, the Cl− can reach the steel bar through these paths, which will lead to the corrosion of steel bar.

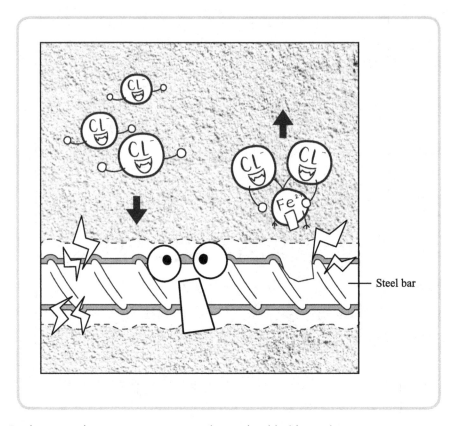

Furthermore, the stray current can accelerate the chloride erosion.

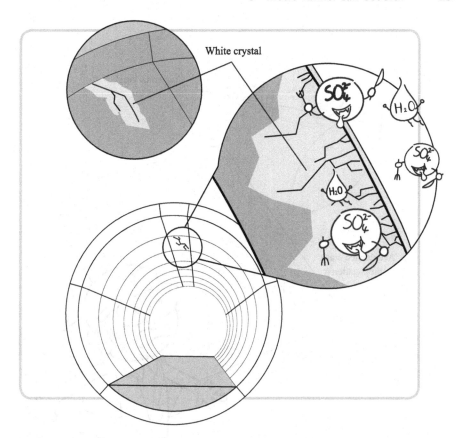

White crystal

The sulfate ion can go through the pore structures of concrete, which makes the concrete to become swollen.

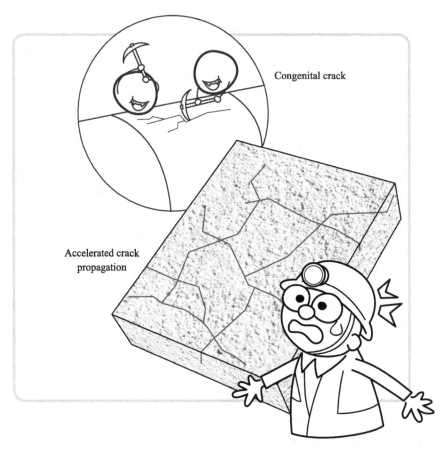

In addition, the existing defect can accelerate the crack propagation. For example, the congenital cracks (usually small) will turn to large cracks due to the existing defects.

The dynamic train can produce the cyclic loading and reduce the fatigue life of the tunnel structures.

4

Digital Information Archive of Metro Tunnel

Is the metro tunnel safe? Each metro tunnel has a digital archive, recording the data related to the life-cycle service of metro tunnel and the digital models under different loading conditions and service conditions. With this health archive, the service condition of tunnel structure is clear. The performance evolution of the tunnel material and the long-term performance development mechanism of the tunnel structure can be better understood. The serviceability of the tunnel structure can be controlled actively and the health service condition is guaranteed.

© Springer Nature Singapore Pte Ltd. and Tongji University Press 2019
H. Zhu, *A Science Comic of Urban Metro Structure*,
https://doi.org/10.1007/978-981-13-0580-1_4

With various types of documents accumulated such as investigation document, design document and construction document, the service data of the tunnel structure could be massive.

Through the Industry Foundation Class (IFC), an ordered digital information archive of metro tunnel is built.

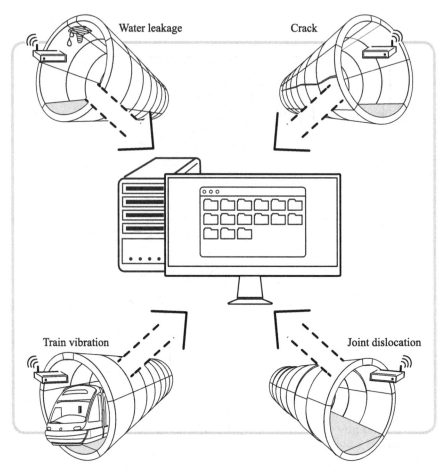

The massive multi-source heterogeneous data including distress data, environment data, and loading data are integrated and managed in the digital system.

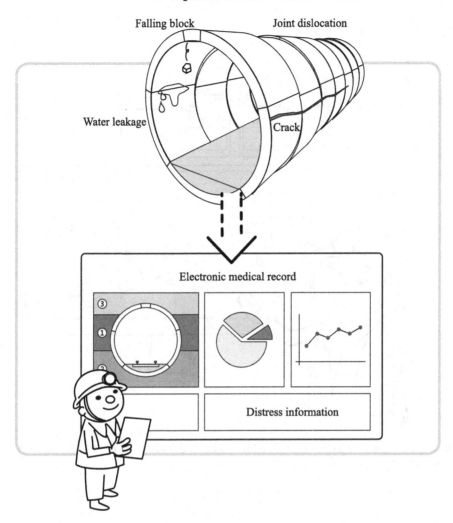

Since the tunnel distress data are stored in the digital archive, the electronic medical record is formed for the tunnel structure service condition.

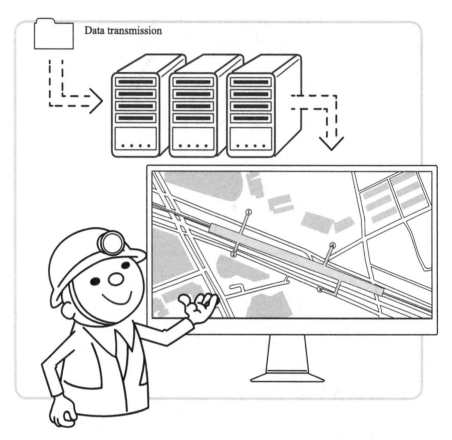

By utilizing the data in the digital archive to build the 2D or 3D model, the structure information is visualized.

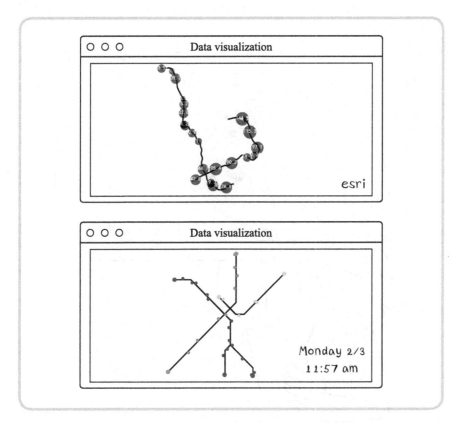

Updating the digital model with newly collected data, the real-time visualization of the data is achieved.

The 3D digital model includes both the geological data and attribute data of the metro tunnel.

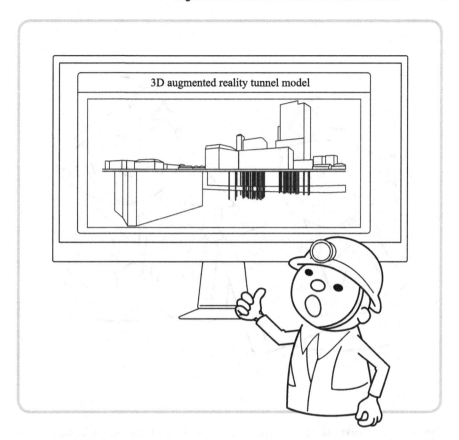

The 3D augmented reality technique can be used to obtain the digital model for the metro tunnel with the real and simulated scenes.

In order to cure sickness of the tunnel structure, it is of significant importance to know the essence of the tunnel structure and material.

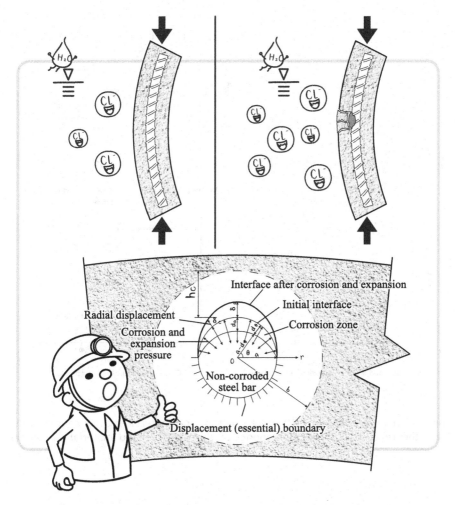

The tunnel is constructed in the underground environment. Therefore, the coupling effect between the environment and tunnel structure should be analyzed.

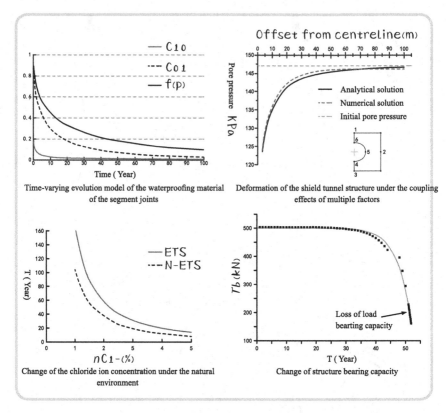

Time-varying evolution model of the waterproofing material of the segment joints

Deformation of the shield tunnel structure under the coupling effects of multiple factors

Change of the chloride ion concentration under the natural environment

Change of structure bearing capacity

The performance evolution of tunnel structure and material is a dynamic process. Thus, the time-varying evolution model of the structure and material should be analyzed.

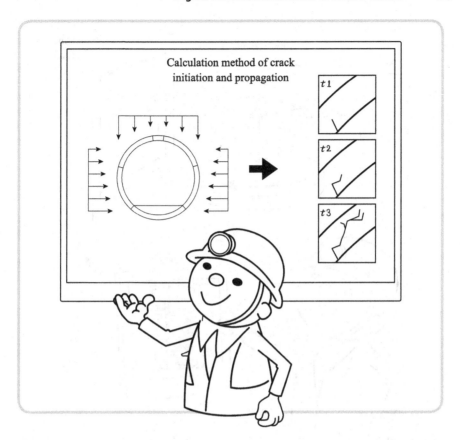

Crack is one of the tunnel distresses. The experiment and numerical simulation can be used to tracking of the evolution of structure cracks.

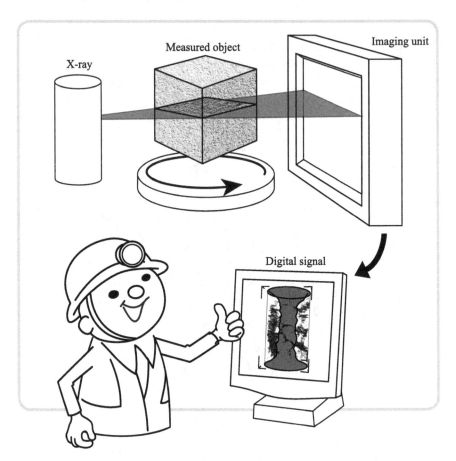

Among the various methods, CT scanning is a perspective way to analyze the evolution of the material.

Chemical titration is another approach to analyze the material evolution. It can be used to determine the degree of corrosion.

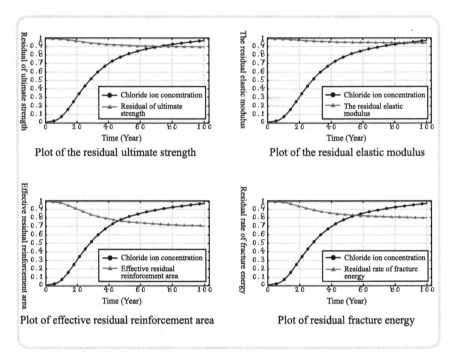

| Plot of the residual ultimate strength | Plot of the residual elastic modulus |
| Plot of effective residual reinforcement area | Plot of residual fracture energy |

Then, the performance evolution of material such as the plot of residual ultimate strength in the above figures can be obtained.

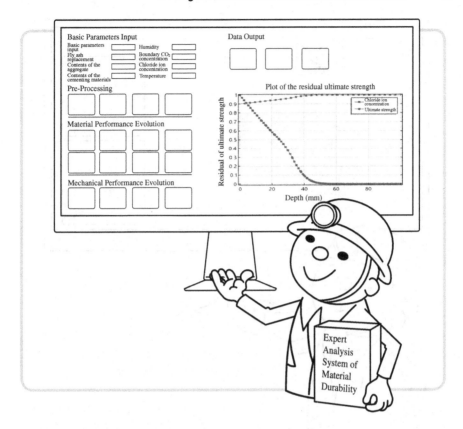

Since the material is in a physical, chemical, and mechanical environment, the performance deterioration mechanism under such a complicated environment can be analyzed.

Based on the integrated digital-numerical platform, the digital-numerical model of the tunnel structural is built to control the service health.

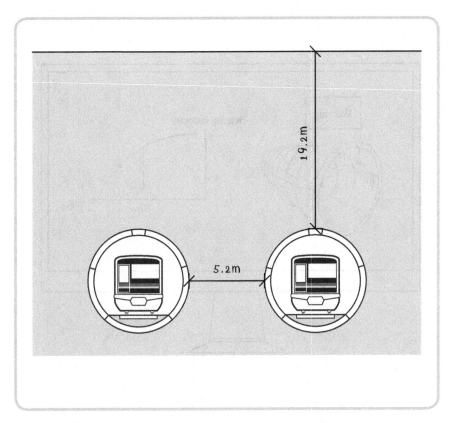

The powerful spatial analysis ability including 2D and 3D spatial analyses are developed in the digital platform to facilitate the professional application and decision-making process.

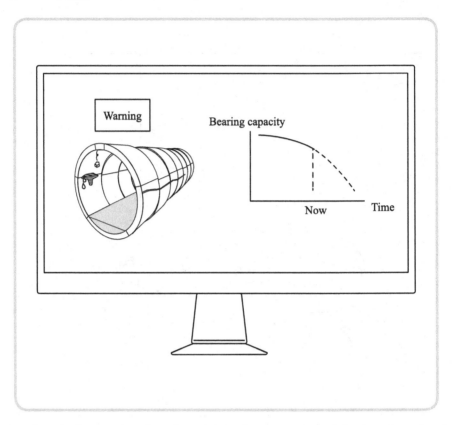

Based on the inspected and monitored data, the structure condition is evaluated and the structure performance against time is predicted.

The repair plans such as aramid fabric and electrochemical repair are analyzed. Then, the decision-making of structural health is conducted.

With the assistance of expert think tank, the multi-objective problems including the aspects of safety, economic, and performance are optimized.

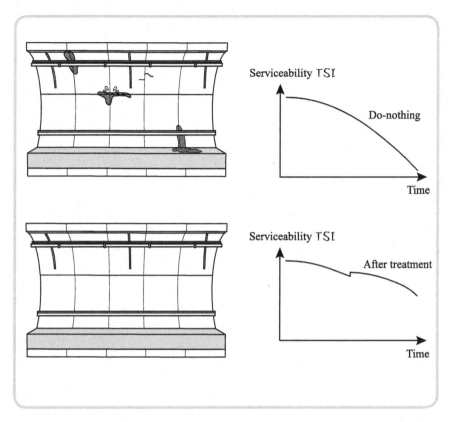

Through the multi-objective optimization, the tunnel serviceability is actively controlled if the treatment is conducted at the optimal time.

5

Experts' Diagnoses for Metro Tunnel

In traditional Chinese medicine (TCM), the doctor can tell the diagnosis after using the four diagnostic methods including observing, listening, asking, and feeling the pulse of the patient. Now the question is how can one judge and tell that if the tunnel structure is in an ill condition or not? The tunnel experts have developed the sensitive and accurate sensors that can perceive the tunnel diseases. Through those temperature, humidity, acceleration, crack, deformation, and water leakage sensors, the values of all types of distress can be measured. The wireless sensor network (WSN) installed on the tunnel structure can collect and transmit the data, so that the distress information can be obtained in time.

In contemporary medicine, doctors have different diagnostic methods to deal with patients in order to find out the real cause of the disease.

In a metro tunnel, the ill symptoms can be perceived through different sensors, such as vibration sensor, crack sensor, and so on.

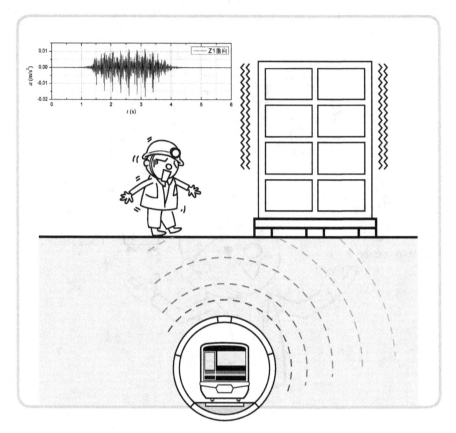

With these different sensors, the tunnel structure is diagnosed fatigue and some measures must be taken to take care of it.

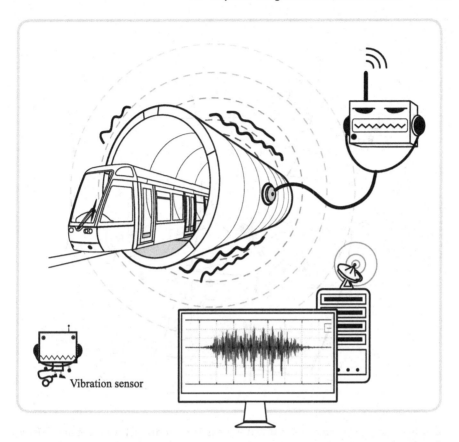

Vibration sensor

When metro vehicle passes through the tunnel, it will cause vibrations and noise. The vibration sensor can identify the noise.

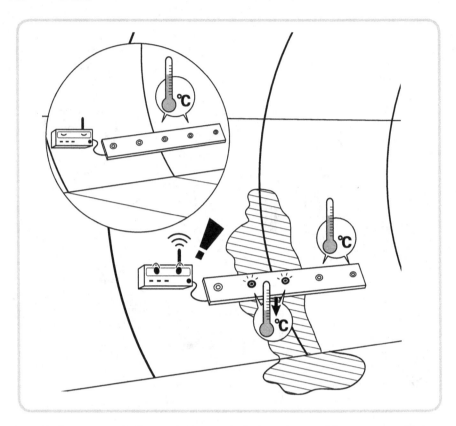

Water leakage sensor indicates that water leaks into the tunnel structure and it issues a warning.

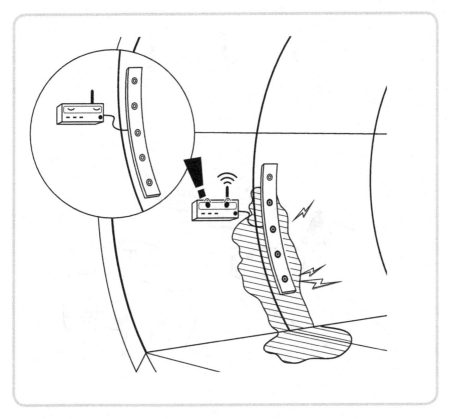

Identification of water leakage by sensors is very important because it indicates which tunnel segment needs to be repaired.

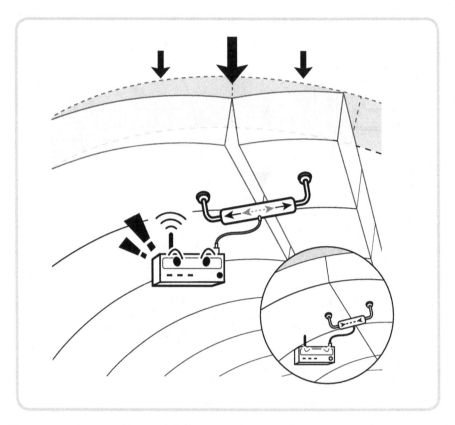

There are defects on the tunnel structure, for example, cracks are often generated during the operation period of metro tunnel.

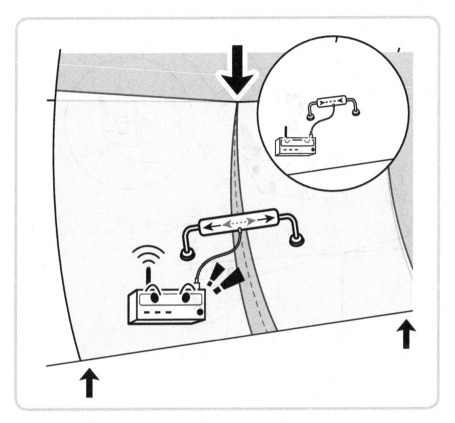

However, the crack sensor can measure the crack. When the crack extends to a larger scale, it will lead to a range of damage.

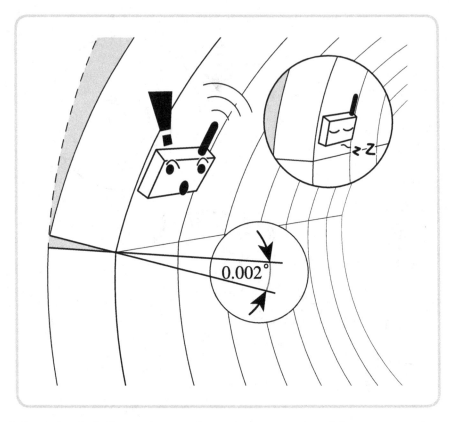

Deformation, including settlement of tunnel segment and angle between tunnel segments, occurs in the tunnel structure.

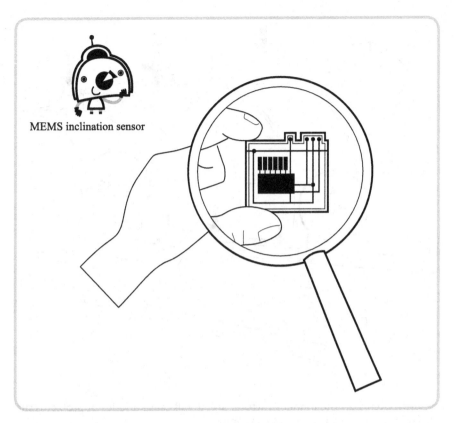

MEMS inclination sensor

MEMS inclination sensor is a newly developing sensor and it can measure the angle between tunnel segments.

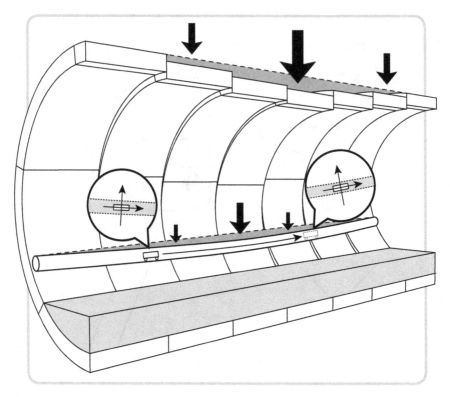

Settlement occurs on the tunnel structure, and it will cause a series of problems such as surface subsidence and segment dislocation.

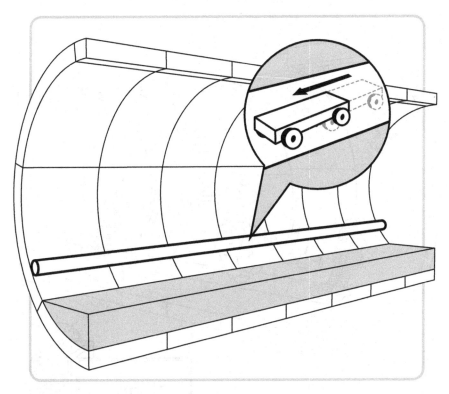

When the smart robot passes through the tunnel structure, it is easy for it to measure the settlement.

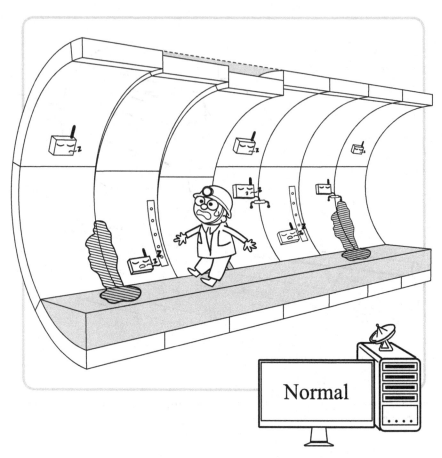

The node layout is very important. If the sensors aren't placed correctly, they will still indicate normal even when the tunnel structure is ill.

The sensors must be installed at important locations in order to perceive the ill symptoms effectively.

The fiber optic sensors are wrapped up in the tunnel and they are just like the bones in our human body.

Every distress can be measured because of the installation of the fiber optic sensors, so the final report can reflect the real condition of the tunnel structure.

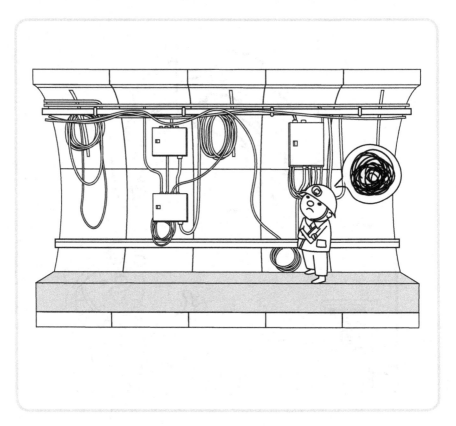

Since there are so many cables placed in the tunnel structure to transfer information, the cable collection is very cumbersome.

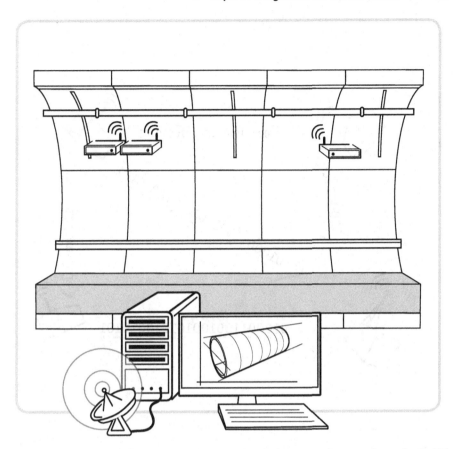

The wireless collection has many advantages, so it is used to replace the cable collection.

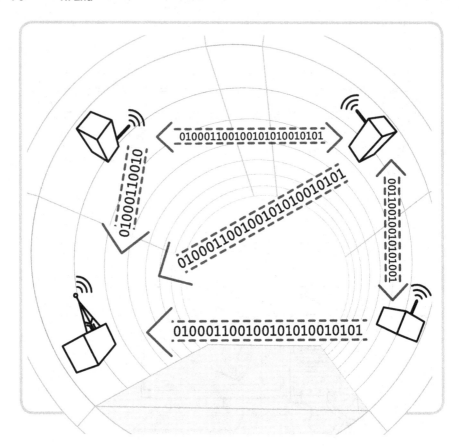

The first advantage is that data compression and recovery are good and data authenticity and reliability are also guaranteed during the transmission.

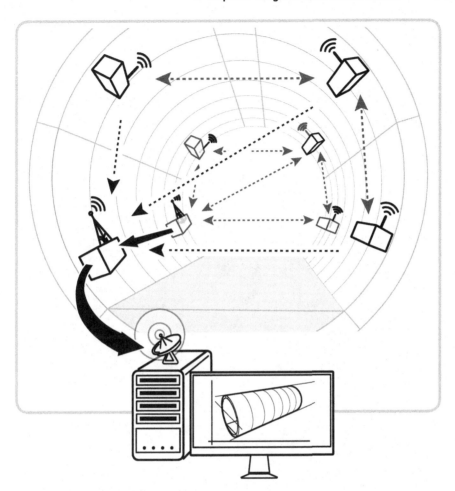

The second advantage is that the power consumption of network transmission is low and it ensures the speed and quality of data transmission.

6

Metro Tunnel Consultation

After the inspection report of the major distress in tunnel is done, the consultation will be held among the experts to determine the current service condition and what's wrong with the tunnel through comparing the distress with the quantitative evaluation criteria and referring to the comprehensive condition evaluation system. The service condition of the defected tunnel structure can be used to predict its condition in the future. The distress development and mechanical changes under the impact of multiple distresses can be predicted.

© Springer Nature Singapore Pte Ltd. and Tongji University Press 2019
H. Zhu, *A Science Comic of Urban Metro Structure*,
https://doi.org/10.1007/978-981-13-0580-1_6

Based on the collected data and information, the experts' consultation can determine the condition.

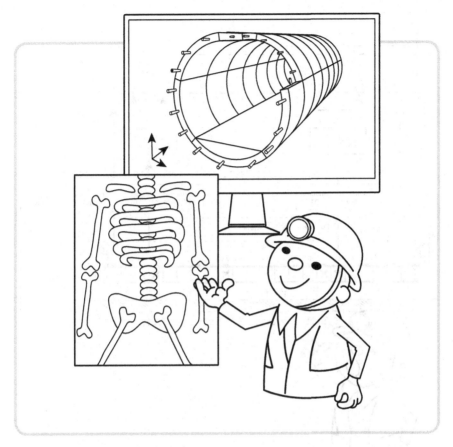

Finite element analysis is an effective solution to mathematical problems, so it can be used to identify the structural damage.

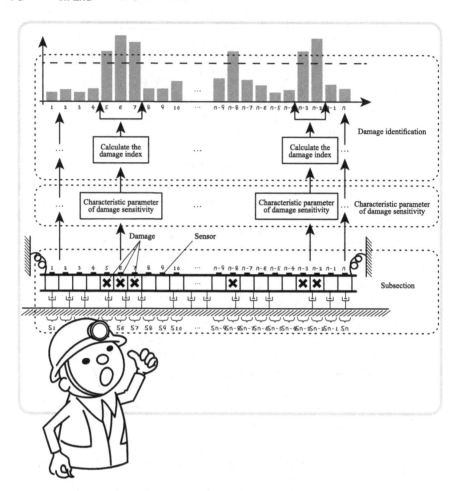

The distributed recognition strategy is a method of solving the problem in a distributed way and it can be used to locate the damage position.

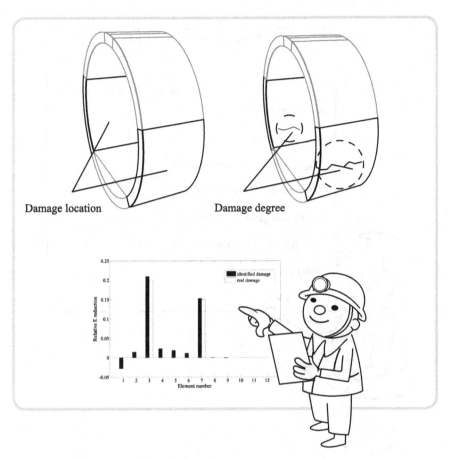

Damage location   Damage degree

The dynamic response refers to the response of a control system and it can be utilized to determine the damage degree.

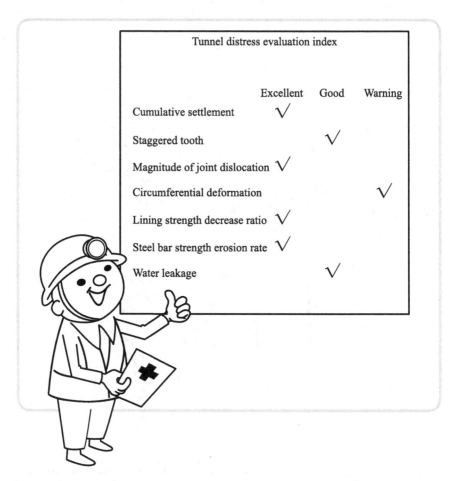

The complete tunnel structure health condition evaluation index includes cumulative settlement, staggered tooth, and so on.

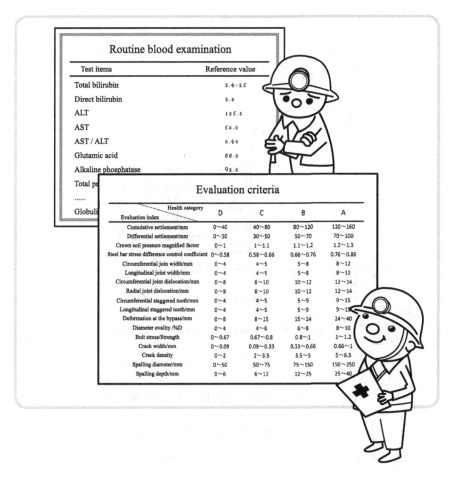

	Routine blood examination	
Test items		Reference value
Total bilirubin		3.4-25
Direct bilirubin		3.0
ALT		125.0
AST		50.0
AST / ALT		0.40
Glutamic acid		66.0
Alkaline phosphatase		92.0
Total pr...		
......		
Globuli...		

Evaluation criteria

Evaluation index \ Health category	D	C	B	A
Cumulative settlement/mm	0~40	40~80	80~120	120~160
Differential settlement/mm	0~30	30~50	50~70	70~100
Crown soil pressure magnified factor	0~1	1~1.1	1.1~1.2	1.2~1.3
Steel bar stress difference control coefficient	0~0.58	0.58~0.66	0.66~0.76	0.76~0.86
Circumferential join width/mm	0~4	4~5	5~8	8~12
Longitudinal joint width/mm	0~4	4~5	5~8	8~12
Circumferential joint dislocation/mm	0~8	8~10	10~12	12~14
Radial joint dislocation/mm	0~8	8~10	10~12	12~14
Circumferential staggered tooth/mm	0~4	4~5	5~9	9~15
Longitudinal staggered tooth/mm	0~4	4~5	5~9	9~15
Deformation at the bypass/mm	0~8	8~15	15~24	24~40
Diameter ovality /%D	0~4	4~6	6~8	8~10
Bolt stress/Strength	0~0.67	0.67~0.8	0.8~1	1~1.2
Crack width/mm	0~0.09	0.09~0.33	0.33~0.66	0.66~1
Crack density	0~2	2~3.5	3.5~5	5~6.5
Spalling diameter/mm	0~50	50~75	75~150	150~250
Spalling depth/mm	0~6	6~12	12~25	25~40

Qualified evaluation system of tunnel distress indices is built up just like the routine blood examination in contemporary medicine.

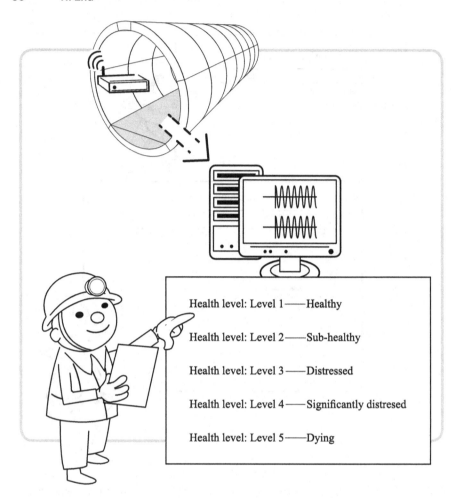

Comprehensive condition evaluation of tunnel structure is divided into five levels: healthy, sub-healthy, distressed, significantly distressed, and dying.

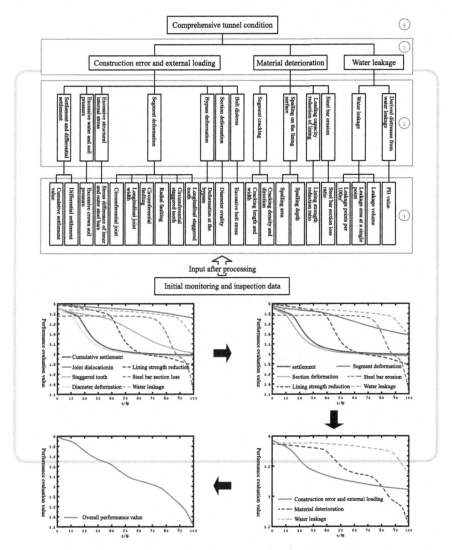

Performance prediction of tunnel structure is carried out through initial monitoring and inspection data under multiple distresses.

7

Metro Tunnel Therapy

The effective way to remedy the disease is to apply the medicine according to the diagnosis. For the metro tunnel, there is a smart self-healing method to treat the tunnel structure damages and an adaptive reinforcement method to solve the problem of tunnel structure performance deterioration. When the damages just occur, the self-healing method can be used. When the deterioration is severe, the adaptive reinforcement should be applied to achieve the active reinforcement.

© Springer Nature Singapore Pte Ltd. and Tongji University Press 2019
H. Zhu, *A Science Comic of Urban Metro Structure*,
https://doi.org/10.1007/978-981-13-0580-1_7

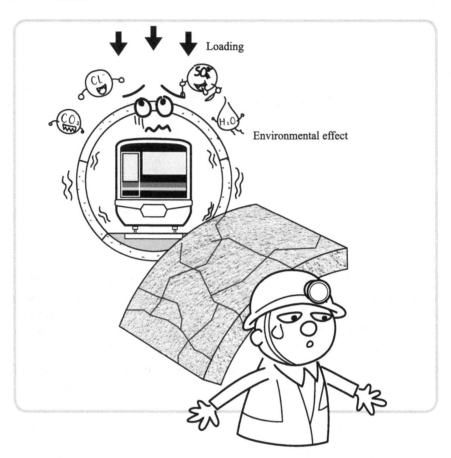

Subjected to complex loadings and environmental effects, the distress of metro tunnel seems inevitable. It is inadvisable to conceal the tunnel distress and take no treatment measures.

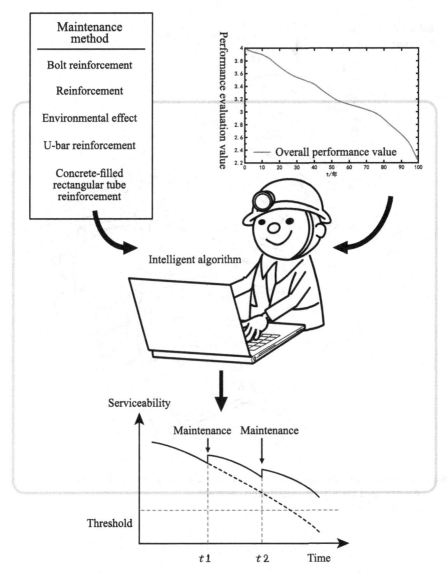

It is advisable to seize the optimal timing of maintenance and choose the corresponding maintenance method.

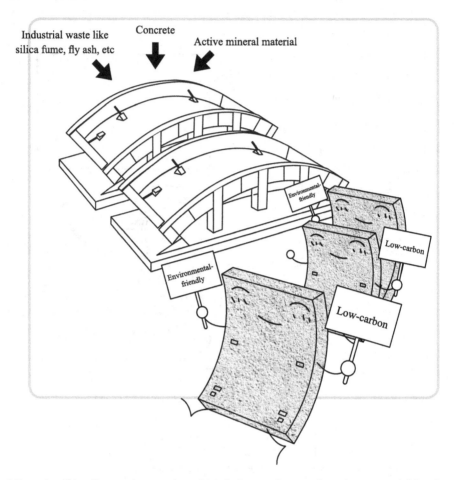

Mineral self-healing maintenance, which is low carbon and environmental friendly, can be employed for newly constructed segment.

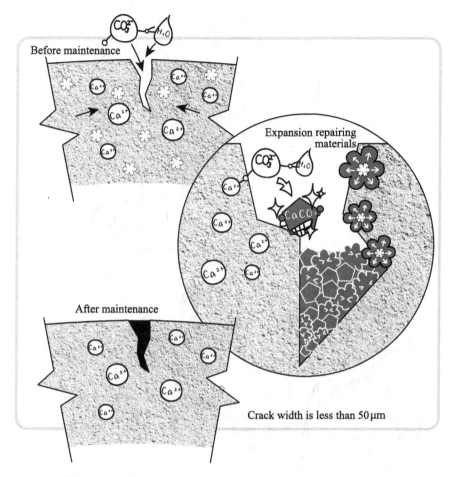

Before maintenance

Expansion repairing materials

After maintenance

Crack width is less than 50 μm

It is compatible to recycle and then turn trash into treasure by expansion component when the crack width is less than 50 μm.

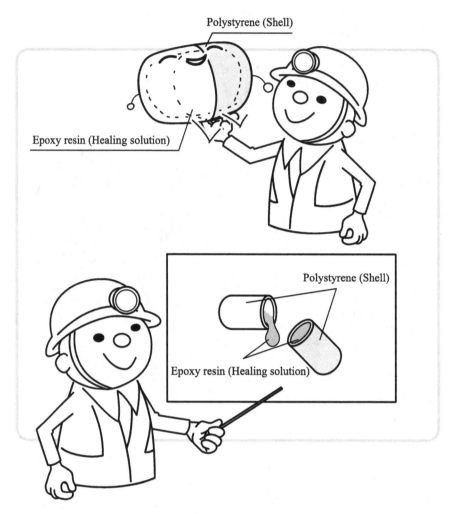

Micro-capsule self-healing with epoxy resin (healing solution) can also be adopted for newly constructed segment.

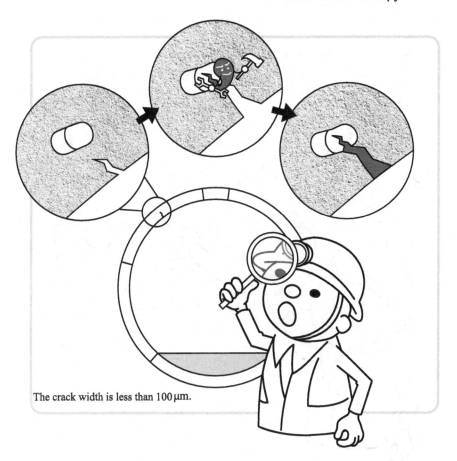

The crack width is less than 100 μm.

The crack is fundamentally repaired by healing solution when the crack width is less than 100 μm.

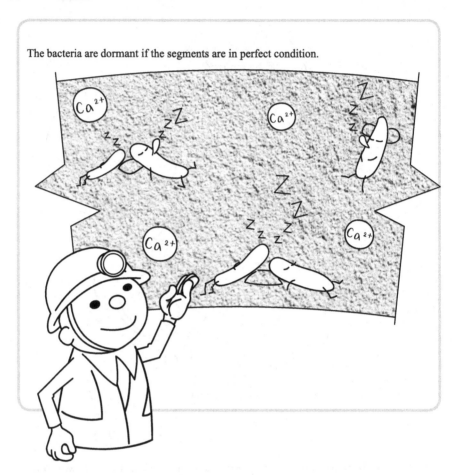

Microorganism self-healing can also be used for newly constructed segment. The bacteria are dormant if the segments are in a perfect condition.

Once the crack is generated, the bacteria trigger the chemical action to generate the substances to fill up the crack.

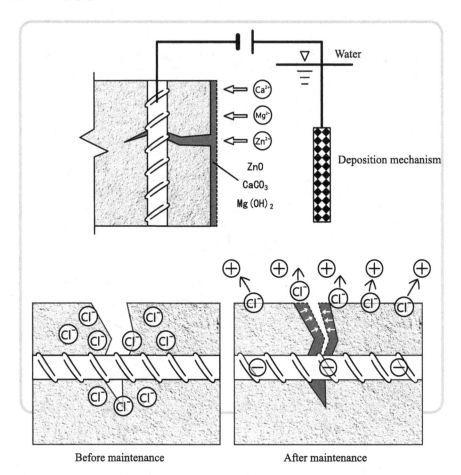

Electrochemical deposition method can be adopted to repair cracks of existing tunnel segment.

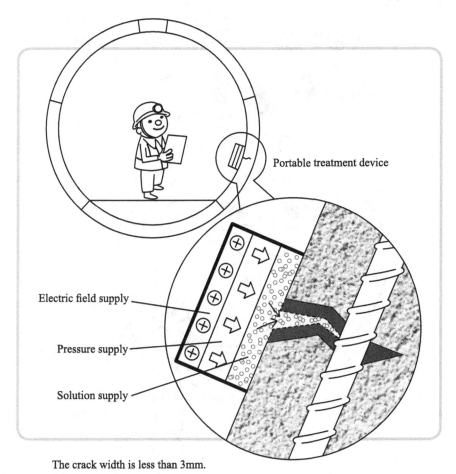

Portable treatment device

Electric field supply

Pressure supply

Solution supply

The crack width is less than 3mm.

Thanks to electrochemical deposition method, the cracks are repaired and the harmful ions are transferred.

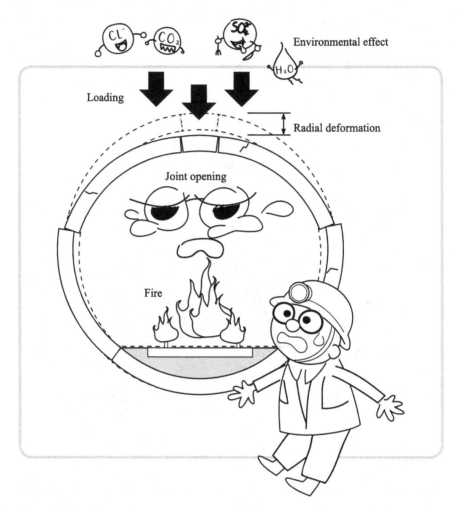

The weakness of tunnel structure should be reinforced, or it will lead to a series of problems including radial deformation and joint opening.

The adaptive reinforcement fixes the weakness and then the tunnel structure is in good health again.

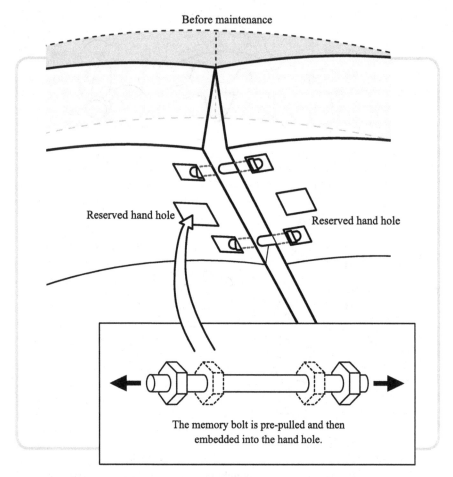

The shape memory alloy bolt, which is pre-pulled and then embedded into the hand hole, is used for adaptive reinforcement in the short term.

After maintenance

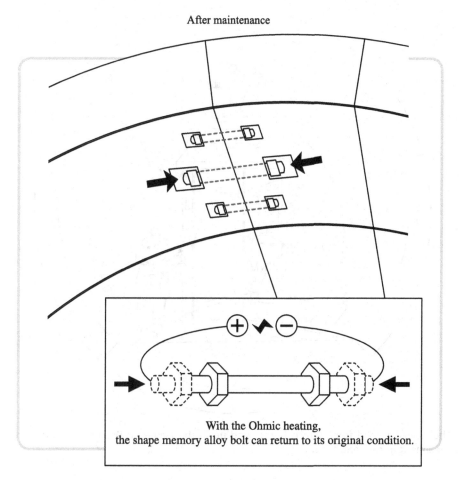

With the Ohmic heating,
the shape memory alloy bolt can return to its original condition.

The bolt returns to its original condition after the Ohmic heating. It is pulled close to the segment so that the joint is closed.

The Kevlar aramid fiber, which is wrapped up in the tunnel, is used for passive reinforcement in the long term.

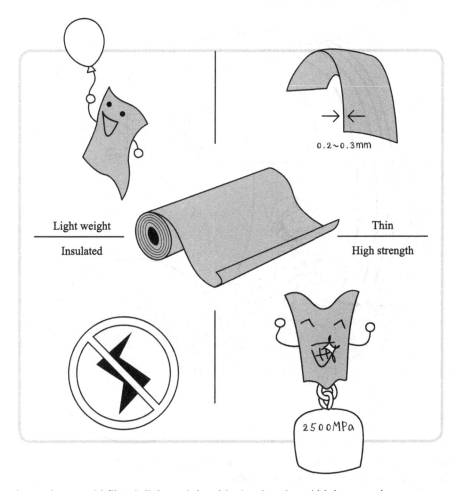

The Kevlar aramid fiber is light weight, thin, insulated, and high strength.

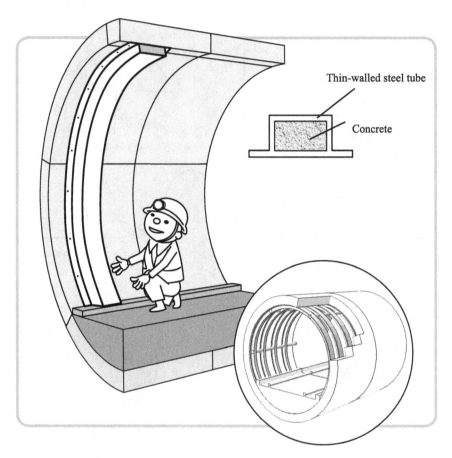

The reinforced concrete ring, which contains thin-walled steel tube and concrete, is applied to bear the loading and deformation.

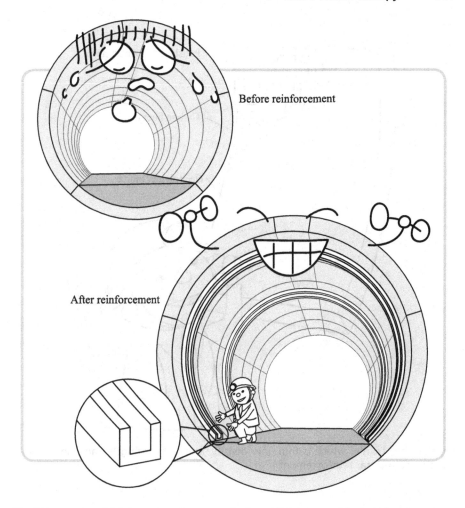

Before reinforcement

After reinforcement

The U-bar arch, which is economic, convenient and light, is used for reinforcement.

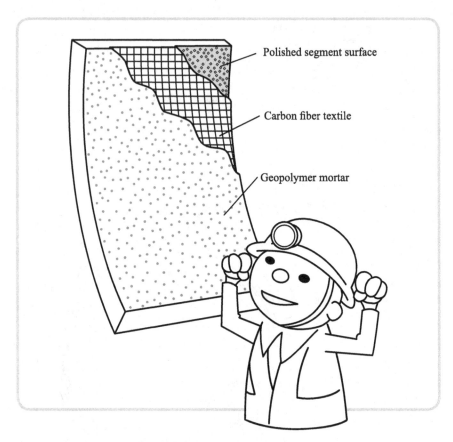

Carbon fiber textile, which is thin, low density, and high strength, can enhance the geopolymer mortar reinforcement.

Both fire prevention and reinforcement should be considered to maintain the tunnel structure in a healthy condition.

8

Metro Tunnel, Aging with Us

As a common commuter transportation vehicle, metro tunnel is also aging old with us like a human being.

© Springer Nature Singapore Pte Ltd. and Tongji University Press 2019
H. Zhu, *A Science Comic of Urban Metro Structure*,
https://doi.org/10.1007/978-981-13-0580-1_8

The life will become better with the reliable metro tunnel.

Printed in the United States
By Bookmasters